THE BEST KEPT SECRET

*"Your Hidden Eternal Identity
& The Most Factual Scientific Reasonings
Which Prove That You Were Born Perfect"*

Eiman "Sheikhy Chic" Makki
**PUBLISHED BY: YOUR WORLD IS YOURS
(YWY) LLC**

Other Published Works, Books & Materials

1. "The Advancing Individual"
   *12 Attitudinal Dispositions
   That Guarantee Personal Advancement" ($19.99)*

2. "How To Not Be Poor"
   Mental Cures Against Poverty, Lack & Scarcity ($19.99)

3. "Believe & Manifest"
   *The Poetically Systematic Science
   Of Belief, Faith & Manifestation ($19.99)*

4. "99 Superior Affirmations"
   *That Will Most Certainly
   Alter The Course Of Your Destiny ($9.99)*

Purchase & Acquire Here

Copyright © 2023 By Eiman "Sheikhy Chic" Makki

All Rights Reserved

TABLE OF CONTENTS

Eye Of God	6
PRELIMANARY NOTES	8
Necessary Scientific Reasonings	9
PREFACE	18
INTRODUCTION	22
PART 1	**26**
CHAPTER 1: "What Is"	27
CHAPTER 2: What You *Is/Are*"	36
CHAPTER 3: The Procession	46
PART 2	**55**
CHAPTER 4: You Were Born Perfect	56
CHAPTER 5: The Roots of Imperfection	64
CHAPTER 6: True (Eternal) Perfection	72

**DEDICATED TO THE SOUL
WHO ALWAYS KNEW THEY WERE WORTHY
BUT DID NOT KNOW THEIR WORTH.**

- *Sheikh Eiman Makki*

Message From The Author.
(Unabridged, Preserved As Originally Written)
June The 10th, 2013 Common Era

Eye Of God

Your World is Yours

In the lives we live often time it is easy to get caught up living day to day not pausing to appreciate what is actually happening. I don't speak about the human relations in which we sulk but about the happening that is. The physical existence the natural forces. This we don't pay attention to because it's normal to us. But I learn to appreciate things, learn to not see things for less or more than it is because that would refer to my value system, which is socially reared, taught during a time where depedance was my status, childhood.

I say all of this to introduce a perspective that I'd like to call the god eye or "eye of god". This is just the perspective that sees what is physically true and not subjectively true. The objective truth. The truth that is true across all subjective viewpoints.

Imagine you were that figure that "created" this existence, what would you value? I'm sure it wouldn't be minor human relations because the way we think amongst ourselves has no true effect on life, just the way lives are lived you feel me? My point just exclaims the hierarchy that perspective has, for example : the ant walks the earth seeing the grass as tall trees the puddle as a massive river & human as a giant. The human sees the ant as tiny, irrelevant, minute and see themselves as normative. We are small in our own sense because if you condescendingly look at our existence in this solar system, this galaxy, this universe we have no impact on the greater picture that is being but like ants we establish our own social hierarchy & assign roles.

But unlike ants to our knowledge we have consciousness which is the key element in living as it gives you choice. As a human among humans the only thing that may hinder you is the roles and social hierarchy but this is not ultimate, the finite relationships between people are not concrete law so don't subject yourself, know your own potential and role in the physical existence and then you see how powerless but powerful you are. With your mind you can put yourself into the situations you desire it just requires seeing things for what they are and moving accordingly.

Just food for thought obtain that Eye,
Your World is Yours
- Sheikh Y.W.Y | Your world is Yours
(June The 10th, 2013 Common Era)

PRELIMANARY NOTES

Necessary Scientific Reasonings

Your World is Yours

The seeds of thought discussed in this document were planted many dates back. Like that of an avocado plant, the sprouting, rooting & fruition has taken its due time, nevertheless, we are here & these thoughts are now. I feel it is incumbent upon me to translate the phenomenon that is *compound comprehension,* that is, the delayed compensation of compiled data gathered through the senses (to say the least).

The ideas/concepts which are urgent in this piece, are centered around the most indirect yet stark, clear, factitious truths there are. Note, by the use of "truths" (as plural), what I as the writer intend to illustrate, or allude to are the innumerable amount of separate & distinct facts which exist, not to give (accredit) power to erroneous & false comprehensions. There is but one truth. One body of truth. Comprised of many smaller truths, which are consistent & are un-conflicting with one another towards the same aim & continual purpose. Moving on, the introduction although dense, serves to warm-up the mind of the reader, so that compound comprehension is achieved through predication based on implications.

It has been made clear through the logics of biology that life, organic material as systems, is fundamentally definable as interrelating organic functions working in harmony to sustain said organic entity/system. If we recognize these facts, alongside the logic that each *"-cosm"* has an adjacent, neighboring -cosm, macro & micro, we can then choose to dial our scope wider or more finely. More *macroscopic* or more *microscopic*.

Let us expand in order to contract, let us breathe life into our minds. If "Life" is "organized organic material functioning harmoniously & systemically", this conclusion naturally leads the inquiring mind to ask "*How?*. By asking the mechanical question of "*how?*", we can then begin conductively reasoning upward or downward, inductively or deductively, toward macro or micro. Let us begin by going micro so we may later go macro, as above, so below.

Starting from the landing (plane) of standing that is Biology, at its essence, let us establish a mutual understanding of Biology & its respective processes. If Biology is predicated upon organized organic material, which systemically interacts with itself to sustain & preserve the integrity of said system, we can now ask, *"What exactly is/are these organic materials? What is organic matter? What is it made of? How does it coalesce?"* Within these questions we now delve into the *logos* of Chemistry. Chemistry, also known during the 8th-14th centuries as Alchemy (*Al-* = the , *Chemy-* = denoting usage of chemicals) is the study of which sub-stances*** constitute, comprise & compound all material. From here let us assume an ontological attitude.

Chemistry is the study of matter on its elemental plane. We know that all things are made of "matter". All things are compositions of particular particles in a wide array of various arrangements. The spectrum of various chemical arrangements, all create the various forms & structures we see in our everyday existence. From the tables, to the walls, paper, pens, dogs, cats, humans & stars. Every object, every *thing* is a compound of matter, substances & chemicals.

With our now established basic scope/view of the primary constituents of material which exists, transitioning into the next basis of our scientific reasoning, we must for the purpose of this document, refine, give shape & coalesce our perspective, not further expand. Before we do so, it is best that we re-cap & synopsize what has already been formulated.

If Biology is the study of organized organic-material, harmoniously systematized to interact so that said entity may sustain, as well as replicate through maintenance of organic integrity, we must make it conscious in the mind of the reader that "Organic" precedes "material". We must take notice that when making mention of Life as organized organic-material, that from here we must infer that if there exists *"organic-matter"* then there must exists other types of matter as well.

Within Biology there exists a slew of in-animate objects, which do not have "life" *per se*. There are many organized chemical substances which are respectively arranged, but they are not *Biological*. This allows us to compound comprehend that there exists *"organic-matter"* as well as *"inorganic-matter"*. The differences are as follows.

1. Organic Matter – Always has a carbon atom
2. Inorganic Matter – Most* do not have carbon atoms (carbon is not ubiquitously sufficient as an indicator for life, so also look for hydrogen, even C-H bonds [carbon-hydrogen bonds are important].

Examples

Organic	Sucrose $C_{12}H_{22}O_{11}$	Benzene C_6H_6	Methane CH_4	Ethanol C_2H_6O	Note Presence of Hydrogen & Carbon
Inorganic	Table Salt NaCL	Carbon Dioxide CO_2	Diamond C	Silver $Ag+$	Note: No Hydrogen Carbon

Now that we fundamentally understand, I'll even say sub-stand, now that we substantially comprehend biology as the science of organized organic matter, which systematically operates in harmony to autonomously preserve its integrity, hopefully long enough to replicate. Now that we can recognize lifeforms as organic matter which by the continual metabolization of other organic compounds perpetuates its existence. Furthermore, also understanding these organized organic material systems (lifeforms) are comprised of chemical substances, interrelating, interacting & expressing, we now have a great stage for the apprehension and perception of Biochemistry, which is a logos unto itself.

For the duration of much of the preceding reasoning, you may have felt an omission of boundaries in terms of organic matter & life, so let it be made clear here, all life is comprised of organic matter, all organic matter isn't alive. When a tree is a tree, it is alive, when it is a table or a door it is still comprised of organic matter, yet is considered non-living. For clarification purposes here are the criterion for "living",

1. Order
2. Sensitivity (Responsiveness to environment)
3. Capacity for reproduction
4. Adaption
5. Growth & development
6. Respiration, Energy Synthesis/Consumption
7. Excretion/Waste Byproducts
8. Movement

Some may be perplexed at the idea of *movement* of a tree, but the process of *growth* of the tree (from perhaps an acorn into an oak), that growth process is the movement. The cellular replication is the movement. So, although it may be rooted & stationary in its own right, it is still living & breathing. Moving.

The purpose of this document is to create frameworks that the mind can further build upon. So, from here we will avoid further diving deeper into nuances of biological & chemical worlds of thought. You now possess an adequate understanding of both, so let us now seek more insight.

So far we have answered *"What is Biology?"*. We have also answered *"How is Biology?"*, which further lead us to inquire into chemistry & what that is. The first landing of standing (plane of understanding) was Biology, from there we assumed a microscopic assessment, which lead us to the scale of chemistry. From chemistry we may delve even more microscopically while at the same time affording us a macroscopic awareness. For our purposes let us use the term *omni-scopic*. The *omni-scopic* perspective will help serve to answer *"How is chemistry?"* while also assisting us in seeing beyond the veil of material as we now move into understanding existence through the lens of physics.

- <u>Physics</u> – *The natural science that studies matter, matters motion & behavior through space & time, as well as related entities of energy & force.*

Hence our usage of the term *omniscopic*. Physics delves into the fundamental minute minutia of all, while still operating on a universally applicable scale, all encompassing. From Biology to Chemistry to Physics we may maintain an inter-logical compounding comprehension. All is all, so we are all the sum of all & are not some of all (partially all), but all of all. We can view & interpret the aggregate value of *all*, without piercing our perception beyond each layer, stratum, cosm or surface.

There are many branches of Physics, as there are many branches of a tree, this is the nature of truth. Stemming from one source, eventually branching off, each branch shares the same essence which is to be found in the root. So, see the many *logos* (bodies/branches of knowledge) as branches which all come from the same tree trunk & ultimately the same roots/source. For the sake of our intended reason-trajectory, we will compound the branches of Physical Chemistry & Chemical Physics.

- <u>Physical Chemistry</u> – *The study of macroscopic & microscopic phenomena in chemical systems in terms of the principles, practices & concepts of physics, such as motion, energy, force, time, thermodynamics, quantum chemistry, chemical equilibria etc.*

- <u>Chemical Physics</u> – *The study of Physio-chemical phenomena using techniques from atomic & molecular physics. Studies chemical processes from the view of physics via condensed matter physics. Example: Quantum mechanical behavior of chemical reactions.*

We have now arrived at a pivotal checkpoint in our journey with this document. We now are linking Biology to Chemistry & further connecting our previous fusion of studies with

Physics, all in order to support the ontological comprehension of our actual condition within the all-encompassing environment which is/are the cosmos, larger than the universe itself.

We can now begin to draw conclusions & with fact based established reasoning rebuke any erroneous conceptions, all predicated upon inferential deductive, inductive & conductive analysis at play.

At this juncture I would like to thank you for investing these moments of your life toward the expansion of your consciousness, you are good to yourself. From the standard plane of recognition, to the micro, to the most macro & most encompassing, the omni-scope, we now have a workable personal view of the various planes of operation which all operates upon. Enough to deduce down to the irrefutable conclusions which are to follow.

We know Biology is the study of "life", that is, organized organic matter which autonomously systematically functions to preserve its individual integrity through harmonious interaction, sustaining itself long enough to replicate by metabolizing other organic materials. Furthermore, we know that all which is studied under Biology, to the exclusion of none, is fundamentally condensed chemical structures of sub-particulate matter arranged & interacting on the physical principles of molecular motion/behaviors. From these conclusions we may further conclude we ourselves are products, active individuations, active agents of all aforementioned processes. Still somewhat vague, we must correlate, corroborate & integrate our inter-logical compound of comprehension.

So far, we have done wonders in differentiating, dissecting & analyzing. Now we must make whole, make holistic the logics, reasonings & thoughts of the materials presented. Using the W5H1 method we may outline many "whats/hows/why's" so that we may further apply to "who/when/where's".

Now that you have been nourished & supplemented with the necessary preliminary information, I can now bestow the elusive, indirect yet stark conception of factual selfhood.

I, Me, You, We are all Bio-chemical reactions of organic matter occupying a vast magnetic field of *Is-ness* & we are all products, individual individuations & active agents in the aforementioned scheme. I will re-literate.

I, Me You, We are all Bio-chemical reactions of organic matter occupying a vast magnetic field that is the *is-ness*, encompassing <u>all</u> within space & time. Not only are we mere products, we are active self-directing agents.

Put this transcription down & walk away in order to give your mind time & space to best allow this new level of comprehension to optimally assimilate into your awareness.

- **Eiman "Sheikh" Makki**

Originally Written by Hand 12/15/2020
Washington D.C., USA

PREFACE

The incipiency of this book was not premeditated. It is not that I, as an author, felt any particular need to write a book on the following subjects. In fact, like most conduits of consciousness, it was almost as if this book came into being aside from the will of my own. The time was approaching Ramadan 2022, which happened to also be the approach of my 29th birthday. To be specific the date was March 23rd 2022, sometime in the late morning.

Prior to the arrival of both occasions, my birthday & Ramadan, I felt a calling and a need to fly transcontinental in order to pay a personal homage to one of the greatest alchemical masters of our modern age, who happened to be my mentor for some time preceding the journey. The teachings ideas and concepts, which will be found in this book, which are to follow, are not of the direct influence from my studies under this alchemical teacher. Still, I felt making the honorable mention that on route to meeting with this individual, these ideas began to form in their incipient stage.

35,000 feet altitude in the sky, sitting abord the airplane, I looked to my left & there was a team of female college athletes. As my mind began to wander & towards less than intellectual or divine spaces, I instantly was overcome with a flush, a flash flood, a deluge of reversed disciplined thinking, which resulted in becoming part one of this book. In all honesty, it was the immediate energetic mental reversal of my base desires, coupled with the authentic sincerity to not even want to resonate with said base desires, coincidently coinciding with the confluence of ideas which I have been meditating and studying for some time, which we may say are responsible for the conditions that brought forth into being what you are about to begin reading.

The greatest flashes of insight, the greatest dawnings of realization, the greatest awakenings, are never those which are preemptive. They are never forced; they simply spring forth into being without the crude & crass forceful expression of personal will. I cannot tell you exactly from whence these ideas came, or why I was chosen to be the messenger. Still, nevertheless, this preface is meant & intended to be a friendly, gentle, warm, solace to the reader that although these are some of the greatest ideas ever compiled through the human thought apparatus (the brain) & it's comprehension, for some reason I feel it is necessary to share the context surrounding its genesis. I feel it is my obligation, as some form of homage or credit, to attribute these ideations themselves to something aside from my personal self.

I consider minds to be like radios, with souls as listeners. Simply, I was mentally tuned to a high enough pitch to clearly receive the transmission at that given intersection of time-space. I had an intuitive sense that being higher in the atmosphere, closer to the radiation of the stars while being away from the electromagnetic congestion on the earth's surface had much to do with the clarity & rapidity by which I transcribed these thoughts from my mind. Just as easily as I could have allowed my mind to spiral down with the frequency of lust and desire, corporeal and carnal gratification, my spiritual discipline, immediately reversed the base-desire-thought-frequency-impulse & sublimated said base impulses into this exact volume. Which we may say, can also serve as a testimony to the potential power of the sublimation of psychical-forces within any given individual. Further expounding upon sublimation as a technique will require a segment of its own, and may appear as a volume in and of itself.

To my teacher at the time, Sir Wes Watson of San Diego, if you ever do read this, know that as we were meeting congregating, getting to know one another, the exact sentiments which are found in this volume were where my mind truly was at the time. I remember glancing at you with the eyes of renewal as I was beginning to realize what we were, individuals yet still it, the living universe. I felt immensely blessed to share experience with you in particular as your presence affirmed much.

May this volume be a testimony to the power of the applied principle of conscience congruency, as I am but a mere transcriber, an obedient servant, to the thoughts which pressed for admission, at that given time. Without congruency in thought, action and energy to my conscience, there would not be this book. Without obedience to the call to transcribe the insight of the moment, the many lives which will be saved, minds which will be cured, souls which will be healed, would not have their salvation, and for this, I must say, without your influence, this book would not be.

I also cannot take much credit for the ideas in and of themselves as they have been explained by many other great minds which have passed through time. Simply, I have done my best to personally transcribe these universal verities into common language. Making available and digestible that which is nearly incomprehensible, inexplicable, and indescribable for many. My mind takes a very technical and analytical approach to very abstract concepts, which is why I find that this book will be of great service to a great many. It is with great honor and joy to present to you all the best kept secret.

-Eiman "Sheikh" Makki
Brooklyn, New York October 11, 2023

INTRODUCTION

Your World is Yours

We must begin by acknowledging that what is true is that the universe & all of the known (as well as unknown) creations have been made. As tautological as the acknowledgement may seem, the realization necessitates the sincerest expounding in order to fully grasp, *apperceive*, the value of said observation. As this manuscript unfolds it will become increasingly clear to any individual that they themselves are *substance* fully fashioned to intercept & inter-ceive itself. It has long dawned upon my awareness that many sincerely fail to recognize their truest self. By "truest self" what is meant is the less obvious intrinsic nature which is eternal, everlasting & which never changes, nor can be changed. In other-words, this "truest self" can also be understood as the inextricable law of being that although sublime, is the most ultimate, infinite & everlasting for all, to the exclusion of none.

Since the moment of my, your or our particular incarnation, each of us have been subjected to being placed under a constant barrage & unrelenting deluge of impressions. These countless unmitigated impressions upon our minds, bodies and spirits in turn have molded, shaped & influenced who & what we perceive ourselves to be. Due to these countless superfluous impressions upon our individual beings, many have developed & embodied an erroneous & false sense of self.

As we further establish increasing realization of the truths that will be later outlined in this manuscript, the common experience of self, i.e., the individual conscious mind, which is the *"I"* in "me" as well as the in "you" will begin to enter its true relative perspective. The self-recognizing part of ourselves which is commonly known as ego, characterized by its ability to establish perception of individuality will in fact come into light as the most ephemeral parts of our mutually respective beings. For many this "I" as distinct from all else is the only basis of identity & as mentioned is ephemeral, transitory, impermanent & as a final foundational bedrock of true reality is ultimately erroneous. The reader who is able to understand will know that the "individual" self is rooted more in transitory realities than not.

It is the relative aim of this writing to embue your consciousness with a new level of self-hood, self-esteem & establish a fact based enlightened personal re-*spect*. In a most analytical, dissective, yet integrative manner, this text will give rise to a converging confluence of ideas, which at their apex will finally serve to illuminate the reader from within. Ultimately delivering an inviolable sense of existential clarity.

Philosophy, Theology, and Ontology Are The Illuminating Sisters Of Fate

شييكي

    This is your initiation. This book is a revelation, transference of un-transcend-able truth. As you absorb, ingest & experience this material, take moments to allow your mind to chew, thoroughly churn, & ruminate in its most literal sense upon this material. For some, the heights of luminosity to be achieved will require more time, while others will achieve illumination instantaneously. The only demand for all initiates is that you allocate sincere, genuine & honest effort to the contemplation/mediation upon these realities. You already know that having knowledge does not confer *knowing*. You must learn on your own. Turnover & make well-rounded contact with this food for mind, food for thought, in order to ensure its proper assimilation into consciousness. Although brief, even the preceding writing above necessitates its respective due time in contemplation to be properly accepted & apperceived.

    You will want to read this piece of writing more times than some, as your current level of maturity of consciousness will contrive meaning only from its current capacity to do so. In the future, as the fruits of wisdom, which stems from maturity gained through experience bloom, revisit this text and surely you will see what you may have failed to previously see. The words locked here in this book will never change, only your comprehensions will, thus the degree of meaning you contrive will evolve. The literal words you see will be the same, but the *in-sights* availed to you by reading through the lens of evolved comprehension will, with time, increase in their potency.

As the author, it is my humble effort to best make use of human words (material tokens of complex emotions) in the attempt to point all seekers of supreme reality toward that which can only be experienced. As you become more proximal to sincere realization, you will recognize that the words only served to lead you to that which is inherently indescribable.

PART 1
"THE ENERGY MADE SELF AWARE"

CHAPTER 1

Your World is Yours

"What Is"

In order to best set & establish an eternal foundation, we must accurately, exactly and most precisely outline what we objectively know concerning the nature of existence from a materialistically-empirical perspective. At the original time of this writing (March the 23rd 2022 C.E) we are well beyond the age of thermonuclear developments and are entirely in a post-atomic age, we may even dare say a meta-atomic era. In the year 1942 C.E., as a result of the "Manhattan Project", western science (more the human species as a body of consciousness) developed the means to split what was at that time the smallest known unit of existential substance, formally known as the *Atom*. The splitting or separation of one of these infinitesimal particles gave rise to the ability to generate unprecedented energy potentials. Sadly, these massive energy potentials were best demonstrated to create damage at scales manifold the weight of the particle itself. Keep this reality in mind as we proceed.

It is known & accepted that I, you, we, that no individual within the universal field can create energy. What is accepted is that energy can only be freed or transferred. This concept is formally known as the "*Law of Conservation of Energy*". It is impossible for anyone to add energy to the universe for they themselves & you yourself are universal energy made material. All the energy which exists is in fact all the energy there is, infinitely indefinite to the human mind, yet certainly definite to that which created the cosmos. It was the freeing of energy which was achieved as a species when the first atom was split & fashioned into the atom bomb. The splitting or separation of an atom is not the creation of energy, only the freeing of energy which already existed.

That was then & now over the almost 9 decades, 80 years and counting since, we have further perfected the release of atomic energy to the point of nearly 500x previous potentials. The applicative efficaciousness & acuity of this power will only increase with time & scientific development. These historical facts have been presented in order to prime and allow your mind to better connect what you already may know with what you are about to further realize.

Yes, most will only recognize the capacity for destruction as the principle value or meaning behind atomic energy, nevertheless, higher minds are able to and will infer more by keeping in mind that material phenomenon are mere symbolical, representational, illusory gateways into metaphysical reality & actuality. Depth of perception beyond surface phenomena is the indicative mark of genuine spiritual vision.

Note what was said was that we split & released the energy from what was then considered the smallest particle (or unit) of existential matter. Key word *existential matter*. This is the foundational base which we will establish our greater understanding upon, in attempt to grasp & realize the following facts

1. Existence exists
2. We know this because existence is happening
3. Above the influence of your will, you & all who you know were made to exist or brought into existence
4. All things which exist are made, created as they have been made, they have been created
5. Things which are, are made of stuff (or some-*thing*) (No *thing* is made of no-*thing*)
6. This stuff of which things are made is what many would call substance & loosely we may understand anything referred to as "substance" as general unspecified matter.

This primary base substance (unspecified material) is existent in all. Again, when we say *stuff*, we mean *substance*, which is best understood as *unspecified matter* in the most general and universal sense. All which is existent is simply *stuff* set into motion, thus bringing forth the phenomenon that is existence. We live in a Universe made of one fundamental substance. The various re-arrangements of said "*stuff*" is responsible for the diversity in the universe. This will be important to consider in later chapters.

"Phenomenal stuff" – Sheikh

The purpose of all physical, material & tangible sciences such as chemistry, biology, and physics are to understand the material world i.e., <u>what the "*stuff*" is, what various "*stuffs*" are, how "*stuff*" behaves etc</u>. Stuff (singular) & Stuffs (plural) may be used interchangeably depending on which strata of material identification one is on & how things happen through various lenses of material interrogation. There is a diverse array of material phenomena. Still, diversity does not render the reality of universality exempt. The one universal truth is diverse in manifestation. The diverse manifestations of *stuff* do not void the singular universality of the fundamental stuff.

Let us take for example a single cell, which is a unit of bio-matter studied in the domain of Biology. If we are to inspect the biological cell further, the school of Chemistry demands that we interrogate the mechanisms of matter beyond the single organic cell & inspect the micro-materials of which a cell is comprised of. Chemistry compartmentalizes material reality on the scale of chemical phenomena. Bio-organic specimen are made of cells,

cells are made of an assortment of chemicals. Physics will further demand that the mind interrogates the chemical units, which are known as elements, and endeavor to understand the realities of said elements. By inspecting an organic cell, we find chemicals & elements. By inspecting the chemicals & elements we arrive at atomic analysis. Cells are bigger structures than chemical elements, and chemical elements are larger structures than atoms. Each are respectively different tiers of material organization. All are encompassed & differ only by stratum of material reality, similar to the ancient Russian Dolls, i.e., microcosm vs macrocosm.

If it would be wise of me, on behalf of the reader, to impart you with a base reference of classification, permitting the imaginative and reasoning mind to conceptualize the imperceptible yet very actual natures of the tenuous fibers of existence. We can start on the plane on which we normally operate on, which is the obvious & visible with the naked eye, from here, further delve into microscopic analysis. Everything which you see around you, is composed of various arrangements of organic & inorganic molecules. Molecules are composed of particles, individual units of elementary matter. Water, H_2O is a molecule, which is comprised of two specific types of particles, yet there are actually three particular particles, namely, two hydrogen particles & one oxygen. This awareness allows us to further seek to conceptualize the microscopic nature within any given particular particle.

If we look at, inspect, the basic elementary forms of matter, any of the many on the periodic table of elements, we see that there are even smaller, even more specialized particles of particulate matter, which we can observe & of which each element on the periodic table is comprised of.

Here we begin to veer into the nature of the *atom*, which was once considered the smallest unit of matter in all of existence. We now know that the atom is comprised of protons, neutrons, and electrons, which by *a-priori* reasoning confirms to us that there are even more microscopic scales of reality and planes of existence, each further encapsulated in the larger macroscopic plane. It was once believed that the atom was the largest until the electron, proton and neutron were discovered. Even further within electrons, protons, neutrons, quantum-mechanics has availed to us that they are even smaller, more infinitesimal particulate forms of matter, namely *quarks*, *leptons* which constitute *fermions*. The visible universe is comprised of *baryonic matter*. Further explanation will require lengthy discourse so we will for the sake of brevity proceed with our intended purposes.

The brief reasoning above allows you to now conceptualize what we mean as *stuff*, more imaginatively, as well as more reasonably than before. As the more microscopic we become, we realize there are infinite degrees, very much beyond human perception & perhaps conception. We don't mean to get into anything further than necessary. What is necessary is for you to be able to effectively conceptualize of that which is meant when we mean *stuff*. You're speaking of the finest of fine substances in all of the universe. The smallest grain of which all matter is comprised of. Let us continue.

In the final analysis, the aim of this text is to aid the student-seeker in their ontological contemplations so that they may develop a more integral awareness of the nuanced interstices of the various sciences & apply said awareness to the bettering of their quality of existence.

Science (or *a science*) can be boldly defined as <u>*any-body of organized facts.*</u> This text is aimed at making the un-transcend-able identity, your un-transcend-able & un-transcend-able truth irrefutably clear. This text will serve to establish your eternal identity. The *raison d'être* of this revealed insight is to make any and all readers aware of their self on one of the highest planes of mental recognition possible so that they may rise above & transcend all transient, limiting subjective identities of which they have been erroneously ascribed or have falsely assumed. <u>*The beginning of all limitation upon ones individual being begins with how one chooses to identify*</u>. The problem we are solving is the raising of the self above subjective identities, which although assumed *real*, are only real through the mind and ultimately are the least useful, pertinent and objective. Your subjective identity will limit you if you let it. Your identity is the limit of your expression. You can only conduct yourself as you believe you ought to based on what you believe yourself to be. This book will teach you to know *what* you are, which is more important than *"who"* you are. What you are is objective & eternal, who you are is transient & tentative.

We will finalize this segment by wrapping up with this series of conclusions & inquiry

Existence is. We ask what is existence? We say occurrence. We ask what is occurring? We know some-thing. What is this some-thing? We know it is a happening of stuff & that this stuff exists. So, we indubitably confirm to know all that exists is made of stuff. It is the purpose of science to specify & categorize said stuff. Science (or a science) is defined as any-body of organized facts.,

Again, "stuff" exists, without "stuff" no-thing would be. It is the various diverse arrangements of basic, primary and fundamental "stuff" which makes all things. We know that we exist. If this were not true you would not be reading. Therefore, we must without a doubt confirm that we are that "stuff", I am that "stuff", you are that "stuff". This book is that stuff, the ink on these pages is that stuff. All is stuff, both universally as well as fundamentally. What is meant is that all is made of the primary substance & that all things are made from "stuff", which again, all things are comprised of.

Various sciences will allow us to move onto other inquiries such as *"when did stuff begin?"*, *"how did stuff begin?"*, *"where did stuff begin?"*, *"how does stuff work?"*. One answer *we may* never know is

"WHY does stuff exist?", one postulate we may put forth is that "stuff exists for the perfection of itself". Still "why does stuff exist?" is a question which even if known changes nothing of the facts, nor will the answer free anyone from the responsibilities of existing. The responsibility of which is to grow & exists according to the laws of your true nature. Which is partly hat you here to discover. Now that we know "<u>what is</u>" let us move specifically on to <u>what you are</u> & <u>how you/we came to be as you/we are.</u>

CHAPTER 2

Your World is Yours

What You *Is/Are"*

What came first the chicken or the egg? As trivial as the question appears, it is a question which even my 13-year old mind knew was something worth its own weight in contemplation. What logically must have come first, had to have been the egg, or at least *an egg*. The question is more loaded than commonly assumed. Knowing that the egg came first does not suffice. Although what first came was the egg, it was not a simply chicken egg. What in actuality came first, was an organism which preceded modern chickens, which happened to also possess the ability to generate/lay eggs. The scientific catalogue of life, the phylogeny affords us the knowledge that most birds type organisms are the closest living descendants of dinosaurs. Specifically Theropods from the Triassic Era 250,000,000 years ago. The ability to generate & lay eggs started ages before the first organism presumable to be a "chicken" even existed.

Aha, so why is this seemingly trivial question pertinent, significant or of any importance for the orderly establishment of our eternal identity? Solving the question of "chicken or egg" is necessary, in order for us to lead with a familiar postulate, so that we may develop and establish a viable-continuum & mental timeline of bio-organic material development. With our logic & understanding now made linear, it is much easier to clearly depict the true evolutionary timeline, so that we may absolve the *almost* platitudinous question of its spurious coloring.

Before we proceed, it is assumable that some may find difficulties in adjusting to the Evolutionary framework. So, it is best we define operationally *"Evolution"*. For our purposes the word, term & phenomenon known as "*evolution*" is simply another word for *unfoldment*.

Most who possess shallow & superficial understandings of sacred texts passed down through the eons i.e., dogmatic consciousness, fail in recognizing that evolution, succinctly, *is the orderly understanding of what "Gods Creation" is & the developmental unfoldment of creation.* Evolution can be deduced to *"the historical development of the known universe"*.

Evolutionary thinking traces millions of years, billions of years, and precedes the formation of the Earth itself. In fact, "*Genesis*", the initial creation story found in the Bible is a figurative conveyance, yet even if looked at literally, the genesis story represents & is the original theory of evolution, preceding Charles Darwin's modern supplementation of Naturalistic explanations.

Secular Evolutionary Theory & what many call "Religious/Theological" creation stories do not conflict. In fact, science & religion compliment & conciliate one another & any consciousness which is truly devoted to the expulsion of biases, erroneous conclusion & the discovery of truth will readily see so. Are art & science truly different at heart?

The discordant sentiments between scientists & religionists are merely byproducts of poor exegetical interpretive skills & stubborn dogmaticism. Only those who lack clarity of sight as well as depth of perception, those who possess poor comprehensive understandings of intermediary facts, wind up cognitively unreconciled, conflicted & discouraged in their attempts to make sense of inter-logical approaches to knowing. Those who fail in finding the intrinsic truth, typically end in disarray & unending arguments.

The purpose of the initial "chicken or egg" question was proposed with the intent to show you that there are various scales & windows of time which differ depending on what degree of inquiry we are examining. What came first the chicken or the egg? This truly may be a window of let us say, 1,000,000 years. What came first the Earth or the elements? The answer to this is a concern of 1,000,000,000's (Billions) of years. Of course elements precede compositions, correct?

In order to develop the best ability to properly visualize, comprehend & interpret time, as it relates to cosmological evolution, multiple dimensional measures of scale/degree are of indispensable importance. This is because all is unfolding & *stuff* is eternally becoming what it will on different scales & planes simultaneously, *co-incidentally*. All products & objects of the cosmic typhoon develop on & across varying scales of time. As we evolve (unfold) from the initial & elementary[1] *stuff* into more specific & complex structures/objects/things, we must know that anything which "stuff/substance" becomes is unique. All things are unique in the time required for their formation, as well as special/environmental conditions needed for their formation on & in the timeline beginning from the initial creationary genesis[2].

"In The Beginning Was The Word (word figurative for impelled vibration) = Big Bang"

So let's refresh

1. Existence *is*.
2. We ask what is existence?
   a. We say occurrence.

3. We ask what is occurring?

---

[1] *note that word *element*-ary
[2] The primary initial impulse which became the universe today

a. We say/know some-thing.
4. What is this some-thing?
   a. We know this "thing" is *stuff* & this *stuff* exists.
5. All that exists is made of *stuff*.
   a. We know that it is the purpose of science to specify, classify & mentally organize said *stuff*.
6. Still never the less, *stuff* exists.
7. Without *stuff* no-thing would be,
8. All the various arrangements of this basic *stuff* make all things.
   a. Therefore we are that *stuff*, I am *stuff*, you are *stuff*, this book is that *stuff*, *stuff is*.
9. Being primary substance, *stuff* is the primary substance that all things are made from, that all things are comprised of.

10. Smaller than an organic cell, smaller than hydrogen, smaller than an atom, smaller than an electron, proton or neutron, smaller than a neutrino, smaller than a quark, so small that we cannot even with the best technology perceive the actuality, but by mentally increasing our conception we may become more intellectually proximal to the nature & reality beyond perception.

11. Even by conceptualizing at our best we still remain at a great distance from the true degree of finitude which in all actuality exists.

Various sciences will allow us to move onto other inquiries such as

1. When did *stuff* begin?
2. How did *stuff* begin?
3. Where did *stuff* begin?
4. How does *stuff* work?
5. Yet one we may never know is "<u>WHY</u> *stuff* exists"
   a. We may say *stuff* exists for its own self-perfection.
   b. Still this is a question which even if known changes nothing of individual immediate duty to the present, that is to say, even if the question of "Why" stuff exists were provided an answer, the answer would not alleviate any individual from the responsibility that existence demands.

When inquiring into these matters, we must understand that from the beginning when *stuff* began existing as we know it to be, *stuff* has been unfolding, that is to say, the primary *stuff* has been developing & has increasingly evolved from elementary things, into various degrees of more complex things. **Yourself, myself, as well as the pages & ink in your hands are included in the list of complex things stuff has become over time.** Reaffirming[3], that all which exists is made of *stuff* & this *stuff* is in all things. Even you & I began as a single sperm cell & now we are individually made of 30,000,000,000,000+ cells, mashallah.

Our new question now becomes, *"what came first, the human or the stuff?"*. The answer is clearly that *stuff is the precursor*, because *stuff* & things other than *"stuff in human form"* have & will exist independent of humanity. This argument, better, this *assertion* is easily accepted because *stuff* has existed before humanity & very well will after humanity. Are you not already beginning to *recognize & realize the power that you <u>are</u> & can never transcend?* The truest, yet almost sublime fact about yourself is that *you are what is & what is has been in existence/previously & prior to yourself, & lastly has never changed in nature, only in form.*

So plainly stated, **you are what is** & although this may sound like "Broken English", <u>*you is what is*</u>, as is, before any false conception of self may be adopted or assumed & you can never not be what is, which also happens to be what was & has always been, understood? All the energy that ever was is still the only energy that is & ever will be because remember, energy cannot be created or destroyed by humans, which means no energy can be added or lost to what already is. Only through a self-assumed false consciousness, i.e., conscious or unconscious volitional dis-identification with your objective being, that is to say, what you are, can & will you through your own power of self-

---

[3] We may even say <u>**con-firming**</u>

recognition delude yourself into false, erroneous self-perceptions.

As already stated, from the time of your personal in-carnation[4], you have been barraged, assailed & deluged by countless sense impressions in the form of people, places, things, acculturations, & societal values, all of which are fundamentally subjective transient "facts[5]" impressed & assumed unto your-"self". Realize that you have been exposed to a bombardment of opinions which will never transcend time like your ultimate, infinite and unconditional nature, which you are now realizing & if re-incarnation is a real phenomenon, sadly the majority may never truly realize for thousands of lifetimes to say the least.

We are almost at the summit of this reading, where by which you will further attain the most functional way to see/identify/recognize yourself moving forward. You are being given the _best kept secret_, which is the self-image & awareness of your intrinsic identity of ultimate-indispensable value. This enlightened perspective is one that enthrones you in the seat of cosmogenic truth, giving you the ability to psychologically operate from your eternal identity, that is if you are able to successfully discard & detach from the false consciousness

---

[4] In-carn-ation (In-_carne_[meat], _to be put into flesh_)
[5] It is wise for us to note that there are Universal & Eternal Facts which are immutable & never changing & there are circumstantial arrangements which transpire & expire, come to learn this well.

which currently has claim on your emotional & sentimental faculties. You must lose your "limited mind" in order to install & make use of your "unlimited mind".

Take time to pause, mentally chew, digest & assimilate the aforementioned facts before reading ahead. Know that those who are eager, who thirst for the wisdom of spirit, mustn't consume so rapaciously that they evoke or invite the discomforts of psycho-emotional indigestion. Know that your eagerness & zeal to acquire & possess this knowledge without approaching with due temperance, may cause you mental indigestion & potentially spiritually discomfort you. So it will require you to re-ingest so that you may properly assimilate said wisdom. I read somewhere once *"A mind once stretched can never return to its original dimensions".* Each stage of enlightenment must be paid for with a penny of cognitive dissonance. *There is no going back.* You have likely made a good deal of haste reading thus far. So re-read what you have read at least 9 times so that you may fully grasp, digest, integrate & assimilate these simple, yet deeply profound mental materials before proceeding. Read for comprehension & not recitation.

"There Is A Profound Difference Between Phonemic Awareness & Reading"

— My Beloved Grandmother

CHAPTER 3

Your World is Yours

The Procession

What came first? As you know by now, what arrived first is *"stuff"*, otherwise known as *substance* or *primary initial material*[6]. Every-thing which exists in the present is made of that same primary existential material. Simply, time has only re-fashioned, re-composed, re-arranged, & assembled in different forms, that is to say, *trans-formed* the primary initial substance. As ego-centered & self-centered as many are (to a spiritually fatal fault), the fact that things have existed before you came into being, is proof enough for our claim. The fact that your parents existed & were before you & before them existed your grandparents, furthermore, at some point neither you or your parents were even thoughts (mental formulations), let alone material formulations, informs us that stuff has been happening before your ability to even conceive of it.

---

[6] Energy & matter are but differing gradations of primary forces, i.e. "stuff"

It is time for you to transcend & lift yourself out of the anthropocentric time-scale of human life. First, envision far back to the "dawn of *stuff*" which as we know it is 13,800,000,000 (that is thirteen-billion, eight-hundred million) years ago & immediately return to yourself in the present. Since the dawn of known existence there has been a steady procession of *stuff*, meaning, an unfolding & becoming of *stuff*. You are that *stuff* & possess the all the same inherent power intrinsic to that original *stuff*. From here we are now better able to sensibly & irrefutably specify exactly what you are. We here assert that you are, by birth-right, able to identify as "*stuff shaped as the human you are*". You are irrefutably *stuff* specifically fashioned into human form, uniquely formed as the in-duplicable human you are[7]. Repeat the following declaration aloud & do not silently do so, use your vocal cords to generate & send very literal vibrations to reverberate through the known universe with power & confidence. Now say: *"I am that original stuff, the original substance of the universe evolved over thirteen-billion, eight-hundred million years into the unique-human form which is now known as & which has been called (insert chosen name). Who I am is subjective, transient & fleeting,* **what** *I am is objective, eternally so. I will through science, love & reasoning learn the laws of my objective being & learn to obey them so that I may develop my most prosperous existence in accordance with immutable law. All laws*

---

[7] Due to your unique genetics, there has never been & will never be an exact formation of you, i.e., another you, you are literally one of a kind from the dawn or time until the end of time.

of my being which I obey will obey me, all laws of my being which I break will break me. I accept the responsibility of my divine inheritance which is the right to become my most-highest potential. I was meant to exist; I was in the imagination of that which imagined all things before most things conceivable ever existed. I was chosen & now I choose my greatest becoming.

We must now ontologically[8] contextualize what all humans are in relation to all *stuff*. Humans are primordial *stuff* manifested into *self-recognizing-stuff*. Our assertion carries a double-edged effect. Humans, are *primordial stuff*, fashioned into a form which is able to "perceive" separateness through their ego, meaning humans with their typically solipsistic[9] minds, may perceive themselves as the center of all experience. Regardless if one chooses to assume themselves as the center of all cosmic phenomena, thinking that the universe is happening *in them*, or if one realizes they are happening in the universe & that they are their own individual creationary center within the Universe, (the latter being most correct), in the transcendent, ultimate & most important sense, *all humans (yourself included) are primordial stuff fashioned into individual beings with the capacity for varying degrees of self-recognition, i.e., consciousness*. This means that above identifying as "John", "Ted", "Lisa", "Abdul", "Sharae" "Xi", you, WE are all individually & universally simply *primordial stuff* which can recognize that it is stuff! Again, personal capacity for varying

---

[8] *Ontology* – Branch of metaphysics dealing with the nature of being, objective/intrinsic realities of a thing.
[9] *Solipsism* – The view or theory that the self is all that is known to exists

degrees of self-recognition is the measure of one's level of consciousness.

Some may find more ease in conceptually framing the point being made as "we are the universe experiencing" itself. Still, this phrase has become a tired cliché, which unlike this reading provides us with no concise, clear or penetrative understanding of what the statement means on the most specific, detailed & psychologically functional level.

You are, we are *stuff* fashioned into a <u>form</u> which we happen to agree to classify as <u>Human</u>. Furthermore, we are *stuff* which can interact with other forms of *stuff*, experience *stuff*, live to know *stuff* & simultaneously be that very *stuff*. You are the divine substance, imagined by the divine imagination, created by the divine will, manifested into a form which can recognize, as well as dynamically experience & intentionally direct its own being, as well as becoming. To me (the messenger) it is deeply concerning (unsettlingly so) to imagine that there are those reading this clear body of facts who will fail to realize or appreciate the power in this book, simply because they are attached to a limited concept of themselves & thus remain buried under needless suppositions imposed upon their *pure consciousness* since the time of their incarnation (what is referred to as "*Fitra* فطرة" in Islam). Many may never find truth simply because they are attached to their *most false self*, afraid of their most true self, the un-transcendable & irrefutable essence they always have been & always will be. The great fear lies in the impoverishment of mind beset by the realization that what one once has believed to be ultimate, was but a transient & unessential conception.

Behold, when they say "you are god" it is not meant that you individually are the sole creator of all *stuff*, but instead that *you are the created stuff, set into motion & given consciousness of itself*. The mutual creator of all creations, that which existed before beginning of space, time & *stuff* (matter/energy) has created all. They say energy cannot be created or destroyed; this is only a *partial* truth. Those who say energy cannot be created or destroyed forget one thing, *energy has been created*. The entire statement which I have personally concluded upon by way of my own *a-priori reasoning* without external influence is *"Energy cannot be created or destroyed by Humans"* or better yet *"Energy cannot be created or destroyed by that which has been created by energy & thus is contingent upon energy"*. We have already made mention that all the energy that ever was still is, this is known as the principle named the law of conservation of energy. The entirety of creation (the universe, multiverses, the cosmos, etc.) has never gained an iota or lost an iota of energy in all of its time of existence. No energy added, no energy subtracted, no energy created, no energy destroyed. All the energy & stuff that ever was, still is & *you are it*.

You are the created stuff with the ability to self-direct & fashion your personal lot of stuff (body, mind & energy) according to your own will & image. Unlike all other creatures

& creations, you as a human being possess more than an autonomic existence. You are gifted the power of "being" what you so will & becoming whatever you so image & choose. You have the only form of celestial body in the entire cosmos[10] which you know can direct/conduct itself according to its own whim & choice[11]. The sun, the moon, the stars, the trees, even the angels according to Islamic folklore have no free-will[12]. Those who are still walking in a somnambulant state merely exist in an unconscious waking slumber & without this level of awareness & awakening. To merely exist is autonomic, yet to *be* requires a conscious effort of the soul. Most are without the slightest awareness of this true, ultimate & objective degree of un-transcend-able knowledge. Those without this awareness are mentally-limited and confined by their ignorance to creature life, compulsively following the whims of organic impulse. Only those who transcend & establish their identity in the foundational yet universal self-conception can & will become the masters of existence. The meek, are not the ones who out of docility submit to humans, bend to the will of "man" or become slavishly sycophantic & obsequious. The meek are those willing to put aside their false ego for their true ego of universal oneness, the consciousness which Jesus "Isa" Christ ﷺ sought to confer, which is the state consciousness known as

---

[10] That we know of…
[11] Even to ones own peril & detriment
[12] Free will can be defined as "Ability to self-select, self-direct & self-govern"

Tawhid[13] in Islam. Did Jesus ﷺ not say *"Ye are gods; and all of you are children of the most-high"* (Psalms 82:6), meaning you, me, we all are emanations & outpressings from the most-high descended through time & formation with the ability to image, will & fashion in the image of thought? Did not the Prophet known as Jesus ﷺ also say "Most assuredly, I say to you, they who believe in me, the works that I do they will do also" (John 14:12), alluding that all mankind possessed the same establish-able abilities, if one truly came into the recognition of their being. The prophet Jesus/Isa ﷺ was attempting to teach you how to save yourselves mentally through correct thinking, not to worship & expect salvation from your own character, conduct & thought.

In the following volume we will uncover exactly what it means to be perfect as well as how you can achieve a perfected practical mastery of selfhood, progressing beyond the understanding of *"stuff"* onto understanding the inner powers hitherto kept latent & dormant, plainly due to simple unawareness. You will now learn about your various inherent faculties, how they work & be provided suggestions as to how one can master being their true, ultimate self.

---

[13] Tawhid – The Islamic term for the Mental Recognition of the Oneness of all Being. The Realization of indivisible nature of "God", What Jesus meant when he said "I & my father are one", acknowledgement that there is no separation between creator & the created only degree, i.e. Christ Consciousness.

What is true is that until this moment you have only been a drifting, wayward, slave to lower tiers identity, limited by what (& who) you erroneously perceived yourself to be. Always yearning, instinctually stretching toward reality, but never finding eternal establishment. You cannot, in fact, no one can ever outperform what they believe themselves to be. <u>One can never outperform their self-image</u>. From here on you are & have been initiated into your grander esoteric reality & truth (*Al-Batinya*[14]). You have irrevocably undergone your divine initiation. The mind once availed & exposed can never be unexposed. That which is seen can never be unseen, that which is spoken can never be unspoken, that which is heard can never be unheard. There is no turning back to your previous state of innocent benightment, you hereby know truth. Prepare yourself & be ready to embrace as well as harness your divine inheritance. From this point forward you will forever be developing as well as perfecting exalted proclivities & divine inclinations.

---

[14] **Batin** (باطن) translates as "Inner", Al-Batinya (باطنية) are the Islamic groups of esoterica

TRUTH BE TOLD

When They Say We Are God, You Are God, I Am God
It Is Best To Ensure The Clearest Understanding
No We Individually Are Not The Creator Of All Creations
We Are Created, Contingent Emanations Of That.
Extensions, Projections, Branches Of A Root.

∞

You Are A Created Creation, Fashioned By The Highest Fashioner With The Ability Fashion Yourself. Allah Is What We Know As The "Creator Of All Creations", Independent And Non-Contingent Upon Creation.

∞

You May Be Deemed A "God" Unto Yourself.
As Above, So Below. As Below, Not So Above.
The Heavens Are Manifested On The Earth, But The Level Of Earth Can Never Be Ascribed To The Manifestations Of The Blessed Heavens

∞

You, I, These Pages Were Laden In The Initial Dusts Of The Created Universe, Borne In Universal Mind Before The Dawn Of Space Energy & Time
– Sheikh "Eiman" Makki

PART 2
"YOU WERE BORN PERFECT"

CHAPTER 4

Your World is Yours

You Were Born Perfect

The reason that you are in possession of this book & are here, now, reading this text, is because it was meant & ordained for you to understand that *you were born perfect*. What is meant by "perfect" is that you were born whole & complete, as an individual masterpiece of the universal creationary order. The elemental, elementary forces which have produced your existence & which have brought you up to this point of being, are an ongoing process of a self-perfecting mechanism which has been working for approximately 13.8 billion years (known to man). The 13.8 billion years of time mentioned, is the agreed upon time which has passed since the inception, genesis & creation of the known universe, the big bang, the beginning of all there is. In other words, what is secularly known as "The Big Bang" occurred approximately 13.8 billion years ago. The forces which are back of the energies & materials of the universe, have been unceasingly self-perfecting ever since. What needs to be understood, grasped & comprehended by the reader is that, when it is said that you are perfect, what is meant is that you have come into existence as a perfect being, with all the materials and faculties necessary to further self-perfect under *self-directed conscientiousness*.

    Let us be very clear, when we say you are "perfect", we mean in the most material sense, as an energetic, organic, physical being, with spiritual faculties as well, that you have been fashioned, formed & created without defect. The "real you", which sits within your vessel/body, has been bestowed & has inherited a perfect vessel. Not only have you been gifted by the creationary order, a perfect machine to operate, you have also been equipped with all the immaterial faculties and tools needed to best grow intentionally, meaning orderly into ever higher forms of personal evolutionary expression. What you must come to learn, is that you must begin to master the use of this pre-perfected sets of resources, endowed to you by the universal creative forces, the same forces which are equally responsible for the entire cosmos. You, as a soul have much growing and learning to do. Know that *only your personal wisdom* is imperfect, but what you are in totality has always been perfect, and is always moving towards a greater state of perfection, i.e., *Evolution*. Evolution in the sense of *unceasing change towards increasingly orderly function*. Your purpose here on earth is to gain clarity, understanding & wisdom so that you may _make best use_ of what is already & has never not been perfect.

If a shrewd comparative look at health & sickness is undertaken, we will understand that health in the simplest sense is a state of harmony and thus perfection. On the other hand, illness/sickness is a state of disharmony & imperfection. The common conception & hegemonic mental assumption is that health (or well-being) is the non-normative state, while illness is regarded as the normative, default state. The flawed frame of mind which perceives health as an anomaly is only the result of an inordinately abundant & egregious surplus of self-misapplication & poor usage of what are already perfect beings. That is to say, due to the high rate of individuals who mis-use (ab-use) themselves, only because they do not know they are already perfect, due to those who use their already perfect being in ways which promote further defect, most assume the flawed conception that health is the rare non-normative commodity.

Simply said, illness is more common & prevalent than health. This is only a byproduct of the wide misapplication of "self", through the inappropriate use of oneself & not because of any inherent or intrinsic design flaw.

What needs to be understood is that, it is never the tool, material or the vessel which is imperfect, but it is the user's use and application which tends to be flawed and ineffective. So, we may say that un-wellness is fostered by individual flawed usage of what is already perfect. It is the ineffectual usage of that which has no flaws which is responsible for producing defects in what was already perfect in the natural state to begin with. Living well is a craft, given the best tools, the user of the tools must perfect their craft, i.e., usage of their tools to the best desired end.

Understand clearly that the high statistical frequency of occurrence of misuse & the frequent ineffective applications of what is naturally a perfect being, gives rise to erroneous psychological precedent/assumptions about the nature of being. This precedent leads to conceptual normalization & mass-desensitization, yet still does not override the principle which is universal, which is that the thing, although mis-used existed as perfect & was only imperfectly used, abused. You were born perfect and are forever perfect, so it will only be through the sustained (or prolonged) ineffective, inappropriate & poor-usage of yourself, which can/will be the cause of any defects or afflictions which arise in your own life.

Health (which is harmony) is the natural state indicative to & corresponding to order, orderly being. One does not have to, nor can they try to be healthy or acquire health, as health is the default state of any un-perverted mechanism. The body is always self-regenerating, self-healing & self-perfecting against individual conscious will. It will & can only be the user through their own selective action (free will), who may introduce affliction, which is existential friction. It is the user who introduces that which creates disharmony, illness, and produces defects in what was already perfect. Only you can introduce unnecessary adversity which create adverse effects & thus perverts what was initially perfect.

This is not to say in all cases one is responsible for the maladies which may afflict them. It is safe & respectable to acknowledge that there are many naturally occurring in-harmonies which may result without one's own personal influence (conscious or unconscious). Still, majority of the unfortunate instances found in most infirmaries & emergency rooms, we will see that people have chosen to thoroughly introduce some element from without themselves, which can be directly correlated & traced to the manifested defects & adverse conditions which in general we call illness. All of these individuals were born perfect & still are perfect, but it is through the habitual volitional, voluntary introduction of that which causes inharmony which is solely responsible for producing all undesired effects. It is the sustained, prolonged, voluntary self-introduction of that which is incompatible with harmony which is the main source of experiential discord, resulting as any defective state, generally understood as illness.

It must be made clear that we are discussing & understanding health not only as a biological or physiological concept, but we are understanding health on the universally applicable plane, understanding health as directly synonymous with harmony, well-being, being well. We understand harmony as the seamless, ceaseless, copacetic, synthesis & cooperation of individual parts toward a singular integrous function.

Let us create an imaginable example. Imagine a grand orchestra. Suppose there are 100 members & that each of the members are playing an instrument respectively. If an extra instrument player is introduced, who has the same instrument as any other member of the orchestra, and this new player chooses to play something totally else, e.g. a note which is not in the score or key, thus introducing by their own choice with an already perfect instrument, they will create disharmony and it will be instantly heard and felt by all. The elements of life constitute the individual sounds which make up the whole symphony of existence. Are we to blame the violin? Or are we to blame the cello? Do we place blame for the inharmonious sound wave introduced into the symphony/orchestra on the flutes or the brass, the instruments themselves? Or do we elucidate ourselves to the root cause of error which is neither the instrument or sound. Truly it is the user, or the player of the instrument who produced the discordant sound. The soul/sole conductor.

You must realize that you have been born perfect as we mentioned earlier, the only thing imperfect is your wisdom. Furthermore, what you must come to realize is that you are endowed with the perfect instrument. The perfect materials and faculties necessary to live a highly expressive evolutionary life. Wherever we see illness instead of health, or see imperfection instead of perfection, we may trace any & all negative effects to the misappropriate usage of that which was already perfect by that which is imperfect, which is the individual, their understanding & their wisdom.

So, we must further inspect and seek to understand on a deeper level why it is that of all who are born perfect, many end up in very evidently inharmonious states & become afflicted by illness either on the physical, mental, emotional or spiritual planes. Furthermore, since we now understand that at the root of any individual imperfection is typically a self-imposed imperfection, we here assert & affirm that at the root of all unnaturally occurring negative & undesirable effects is the conscious or unconscious willful promotion of these defects created by the individual introducing some incompatible element or variable unto themselves. As we learn to live more effectively, we must recognize that choices incompatible with the creationary forces (harmonies of life) reproduce inharmonious (destructive) experiences for the individual.

Again, you were born perfect, you still are perfect, and it is in your universal cosmological nature to be self-perfecting. Your natural instinct is self-refinement, self-perfection & self-unfoldment. Strenuous effort is unnatural, sincere effort which treads the line of grace is what is best. Harmony (which is health) cannot be necessarily "acquired" through direct action, instead health is the naturally resulting & occurring state of being, given that there are no inharmonious elements, variables, or factors introduced to that which was already perfect & harmonious. You cannot acquire health; yet one may very well acquire sickness. Do you understand how this works?

Health, harmony & perfection are your natural states & orders of being. Health, harmony & perfection are not oddities which are to only be experienced by a scarce few. Illness is the product of ignorance. In fact, all existential afflictions are the products of ignorance to the laws of being. It can only be through volitional introduction of unnecessary elements, factors or components from without, by the self, which results in illness, injury, imperfection and defects in the life of the individual. Much of what has been said may seem to be circuitous, this is because the realization needs to be consistently revisited so that it is genuinely emotionally comprehended & understood. As we continue to unfold, you will notice the development & refinement of your wisdom. The refinement of wisdom will be the only necessary undertaking to best live well. You are perfect, you were born perfect, you have always been perfect & it has only been through the conscious (or unconscious) willful adding of injurious, inharmonious & incompatible elements unto yourself, which has produced any and all negative effects or defects to your constitution, which again, was initially already perfect. Perfect your wisdom in order so that you may best use, serve & protect you as well as your perfect being.

CHAPTER 5

Your World is Yours

The Roots of Imperfection

Assuming you have read this book in its due order, you now have an exact understanding of how you are already perfect, always have been so and will always be so. We must now further develop your understanding of how, why and for what underlying reasons the majority of individuals willingly add what they instinctually know promotes defects into their naturally perfect constitution unto themselves. We must consciously ask, as well as address what we intuitively ask ourselves, which is *"Why is it that anyone would add unto themselves anything which produces illness, inharmony or defects?* Why would anyone willfully *add* anything deleterious, harmful or injurious to that which is already perfect? The shortest answer again is, because of individual ignorance to one's true nature. By developing our understanding along the lines of these questions, we will come to know the subtle yet *veritable* reason why many perpetually exist sub-optimally.

By sub-optimally, what is meant is <u>as the less than the perfect self</u> that they were/are intended to be and still are fully capable of being/becoming. You ask, *"how can one still be perfect after introducing defects?"*. The reason is because they are perfect, you are perfect. They have always been perfect. You have always been perfect & it is in the natural design & constitution of yourself, beyond your persona, will and comprehension to be a self-perfecting entity. If one were to cover a diamond in manure, the diamond itself never loses its state of perfection. The diamond never loses its intrinsic value. The diamond simply becomes buried under that which when removed restores and reveals the natural state of perfection. So why/how does the mass of souls manage to self-impose imperfections in their own lives?

It isn't that you or many others consciously desire to create imperfections in their lives, neither is it that anyone actually intentionally seeks to ruin themselves. As mentioned in our first book "The Advancing Individual"

> *"If The First Rule And Law Of Nature*
> *Be The Law Of Self-Preservation,*
> *Then The First Crime Of Nature*
> *Should Be Self-Degradation*
> *Otherwise Known As Self-Destruction"*
>
> *(Chapter 8 Final Paragraph).*

It is wise for us to realize that, intrinsically, what many are unconsciously searching for is a betterment of their existential condition. In the quest for improved experience, simply you, they are (or have been) mistaken as to the proper means of attainment of existential betterment. Many seek fulfilment in the same places they discover hollowness. Many seek strength in same places where they find weakness. Many seek ability, where they consistently discover debilitation. Be aware that the only time any individual is spurred towards any acquisition, or consumes anything, is when there occurs the instance of a *perceived benefit*. *Perceived benefit* does not mean or allow us to logically infer *actual benefit*. Take the case of the narcotics addict. The narcotics user may *perceive* that they are benefiting from whatever substance they consume, but this myopic conceptualization of "benefit" inevitably & infallibly proves severely detrimental, even fatal in the long-term.

Again, what we are addressing & seeking to make sense of here is *"Why would anyone introduce that which causes defects to that which is already perfect?"*. The answer is because as a precursor, they falsely perceive some defects within themselves & are seeking to re-establish perfection. Think for a moment, there would & could be no perceived need or desire for remedy if there were no perceived presence of malady or lack. We can never have a desire for that which we feel no need for. Right?

Now that we recognize that it is the false perception/assumption that one is imperfect, which acts as the pre-condition to an individual seeking a solution, by adding unto the self some form of exogenous matter, we may penetrate even deeper beyond the veil into the heart of the matters we are concerned with.

We find that only in times of perceived weakness or perceived imperfection that many seek aid, seek assistance, seek for consolation in materials which are outside of themselves. What is the truest fact is that there exists only true one solution, which is to be found through cultivating internal fortitude & the building up of one's inner-faculties. The only answer to all existential affliction is to begin perfecting the personal use (self-application) of the material that one already *is*. That is to say, making better use of oneself as a means to the ends of increased quality of existential experience (inwardly or outwardly).

Here we can assert that individuals compromise themselves in attempts to seek power/strength from without. But there is always an incurred penalty & necessary retribution demanded for any cheating of the law of karma. If faced with wisdom, every experience opportunes each soul for the means of increased spiritual skill, that is to say internal adjustment to external pressures, which in the end result in *power over circumstances*.

Each student of life must become aware that the law of equivalent exchange will always take & demand a toll for any karmic work shirked in the universe. Look at the many who have avoided becoming spiritually strong by depending on a narcotic

& thus with the passing of time have become increasingly weaker than ever before, less perfect than intended.

By default, as an already perfectly designed being, which has been cosmically engineered to increasingly unfold into higher degrees of perfection, you must _realize all strength, power & development must ensue from you_. Strength is developed, never bestowed. Succinctly defined, strength is the measure _of increased internal adjustment to external pressures_. Pressure precedes the development of all strengths & this is why strength can be best defined as "ones degree of withstanding against external pressure". The pressures of life are meant as aids for each individual to use as instruments of greater adjustment, greater strength and an increased power over circumstance.

The grandest pitfall is when any individual habituates (_inclines_) themselves to irresponsibly deferring any personal maladjustment to an external source.

Herein exists the reason why many are self-defecting & self-compromising. In an attempt to avoid the efforts demanded by karma, which would result in increased internal adjustment to circumstances, individuals such as narcotics addicts defer their pressure relief work to that which costs personal harm & injury. Both manifest & latent[15]. By avoiding the dharma of their karmic experiences, which is meant as the instrument for greater

---

[15] Referring to Manifest & Latent Effects. _Manifest Effects_ being the immediate consequential effects produced, _Latent Effects_ being the unconscious & delayed effects brought about from an action.

strength, the individual sidesteps & thus stunts their own potential for power & growth.

In many cases when one hears "pressures", many are apt to exclusively assume physical pressures, so let it be made clear here that we are making use of the term "*pressure*" as the term pertains to the entire spectrum of subjective experience. Gravity is what is responsible for causing physical pressure. What you (the student) must be cognizant of, is that, individuals who are seeking external refuge & thus compromise themselves, are seeking relief from psychological, emotional & spiritual pressures. These three types of immaterial pressures, are more finer & subtle energetic forces, which although less crass, are higher in vibration thus requiring greater resilience to embrace, withstand & sustain.

Mental adjustment & mental strength development is an absolute essential for living well. You are to transcend the "pursuit" of happiness & seek *eudaimonia*, which we may say may never be acquired via pursuit, but instead must ensue from you. <u>Perpetual infinite emotional adjustment to circumstances by the upbuilding of character is what most needs attention</u>. The inability to psycho-spiritually cope, adjust & adapt[16] to the confluence of perceptions & impression upon ones individual being is what leads many to self-defecting habits.

---

[16] Here what is meant by coping is tolerance for a period of time, distinct from adapting which is an established growth over circumstances which lasts forever.

In an attempt to seek strength, the many only discover & amplify weakness. Where many seek power from without, they curate & promote weakness within. As mentioned, another way of understanding the state of perfection is as the state of harmony. Where many seek to induce harmony, they only promote & generate more disharmony in their lives.

The universal intrinsic desire is always for strength, power, perfection, *well-being* & harmony. The means & methods for attainment are what must be corrected. Until the individual is ready to rise to their own power & realize that they are perfect & have already been perfect, they will continue to seek harmony from without. When the soul matures & benightment no longer prevails, when the soul awakens to the folly of its former ways, the individual soul will cease to add that which has been & always was unnecessary.

It has always been through what was added that any affliction increased, proliferated & multiplied. So, it can and will be by removing the superfluous & beginning to rely on one's ability to meet the gamut of subjectively experienceable pressure with consciously effortful development that all souls will reverse the flaw & thus return to the path of self-perfection.

You were never not perfect & have always been perfect. In the times where you have experienced subjective pressures the choice was always to grow against the pressure from within or refer/defer for strength from without. Those who have withstood have become the overcoming souls, those who caved perished. You were born perfect, still are perfect & are unceasingly inclined toward perfection. The one and only thing about you which has been imperfect is your personal wisdom (understanding). *Experience is the indispensable alchemical resource & material of all souls, meant to be transmuted into the refinement of ones upward being.* All experiences are opportunities for finer degrees of soul fortification.

CHAPTER 6

Your World is Yours

True (Eternal) Perfection

As an ever-ascending student of life, as an always-aspiring & evolving being, you must begin to fully understand, fully comprehend the true nature of the ideal which we call "perfection". The common, conversational understanding of "perfection" has led most of the world to denounce the possibility of the attainment of perfection. The common conception of perfection garners the space to reject any possible existence of perfection. This mass-misinterpretation, in turn, garners more room for perfection to be more infrequently attained. We may hold the lack of *sincere understanding* of the true nature of perfection responsible for the wide-scale unmanifested potential which we see all about us in our temporal reality.

The simplest & common statement that we find surrounding the notion of perfection is that "*nothing is perfect*".

This pithy remark then gives rise to the erroneous assumption & false conclusion that perfection does not & cannot exist. The only individuals who would make such a claim, are those who obviously do not have a clear understanding and/or conception of what it authentically means to be perfect. Those who assert that "nothing is perfect" lack clear, rational insight toward the ideal known as *perfection*. On the other hand, by the end of this chapter, you (the student) will have a more grand, clear, concise, applicable, practical, accurate & correct way of understanding, approaching & aspiring towards the ideal that is *perfection*.

Let us begin with the pretense "*nothing is perfect*". We must assert, that while this is "relatively true", it is not wholly true. First, it is best for us to clarify the operative definition of perfect, as found in the dictionary. Perfect is defined as "*having all required or desirable elements, qualities or characteristics, as good as is possible to be*". Let us keep that last fraction of a sentence in the scope of our minds as we continue to further upgrade, develop and evolve our understanding of perfection, so that we may make better use of the concept-ideal in our own personal lives.

The common majority will say that perfect does not exist and that nothing is perfect. But this only tends to be a subversive way for one to lessen their degree of accountability to known & possible higher performative standards, that is to say, to ease the incumbency upon the individual, to aspire toward greater & higher expressions of evolutionary selfhood.

The key segment in the operatively defined definition which must be held on to is *"as possible to be"*. Once we recognize that we may aspire to a higher degree, while not only recognizing, but actually proactively aspiring to said ideal of which we know is possible to be, we may in fact, refer to ourselves as presently perfect, already perfect in the present. What is true is that one must in the very least, be able to conceive of a higher degree of possibility for themselves, so that they may eventually live up to operating at the level which if recognized is actually possible.

Furthermore, although some may never be in touch with the true actuality of possibility, even those who are in a subjective reality composed of & confined to their own limited ideas of possibility, are in fact perfect to their known degree. As perfect as it is possible to be for their present level of awareness. So, it will be by widening one's conception of possibility, that is, by broadening one's conception of what is possible and then living up to said possible standard, which will best define one's individual degree of subjective perfection.

So, in fact, perfect does exist. Perfection is attainable & there are more inconspicuous nuances for us to discuss, in order to fully free ourselves from the still unresolved instinctually, not consciously held questions, as well as mysteries, still surrounding the ideal of perfection.

Some even suppose that no one is perfect, only God is perfect. Due to the fact that the word "God" has many interpretations[17], we will clearly state an operative definition coined by the "Your world is Your School For Applied Enlightenment". We will here define & agree to use the word "God" to mean *"The Mutual Creator Of All Creations"*. Fair?

If we cosmogenicaly trace back the unfoldment of creation as we know it, we must understand that all of which is, *is creation*. We must acknowledge that there is a source of all creation, which transcends creation, which is responsible for all of that which is created.

What the novice, intermediate or adept must always recognize and avoid is that they must never project an anthropomorphic conceptualization of the *"Mutual Creator Of All Creations"*. For the creator all creations precedes all creation, which by default includes all anthropological manifestations, that is to say humans. Think, the Earth was formed (barren & without life) 4.6 Billion Years Ago. Think some more, the known Universe came into being 13.8 billion years ago. This means that there were 9.8 billion years of time where by which, the earth itself was non-existent, let alone bi-pedal organisms which we call "human".

The attempt to reduce the incomprehensible (or better yet the pre-comprehensible[18]) to conceptualizations limited by the limited detective technology of human corporeal mind[19], would

---

[17] Polysemy: The co-existence of many possible meanings for a word or phrase.
[18] That which existed well before human comprehension did.
[19] The organic senses of perception (eyes, skin, ears, taste, nose)

be absolutely blasphemous in & of itself. Let us continue now that we have one of the best operative definitions of that which is "God".

The benighted many will always say that no one is perfect, only God is perfect, & this may very well be true. The one that is most complete, is the creator of all creations themselves[20]. Still knowing that God is most complete does not confer consent for you to be less than *your known best*, your absolute best. Acknowledging that God is most perfect is not meant to dissuade you from rising to your highest ideal, your highest known greatness.

What you must grasp is that perfection is everlastingly-perpetual & that to be perfect, is a state of being in that everlasting state of perfection. What you must recognize is that, although you may not currently be the absolute best you know it is possible for you to be, you must recognize yourself as <u>*perfection in progress*</u>. That is to say, the key to understanding yourself is that you are <u>*"ever perfecting, never perfect"*</u>. So, in fact, to be perfect, is in itself, a state of being aligned with the highest desirable qualities & or traits, <u>*as best as possible, to be.*</u>

"Perfecting, never perfect", backs ones disposition keeping an open space for continual perfection. That is to say, the one who is perfect is that individual who is eternally progressing.

---

[20] Not to be taken as the creator of all creation being pluralized. Using they/them in an effort to not assign gender, which is a inextricable conceptual artifact of anthropomorphic physiology, namely genitalia.

What you may not have fully recognized hitherto this moment in your life is that you exist in a field of creation. You may not have recognized that you exist, created not by your own personal volition or conduct, and that all the things about you (which are also creations) are just as equally created & existent as you are.

The folly of the immature spirit & the unenlightened (the benighted), lies in their self-perceived separateness from the whole of existence. This immature, underdeveloped, and unevolved ego, that is to say sense of self, which is most prevalent, is most responsible for the lack of auspicious living, and the increase of personal suffering.

The secondary phase of ego which although may be slightly evolved, yet still is not the highest, which precedes the level you are about to be initiated into, is the ego which identifies themselves as existing in a cosmos higher than themselves. That is to say, those who acknowledge that the universe *is*. Still, this second stage, yet not fully evolved, mature level of ego identification is limited, since the individual still fails to recognize that they are still operating from a degree of perceived-separateness. *I-thou.*

Bringing everything back to your new conception of perfection, the truth of the matter is that you are in fact, just as much the universe, as the universe outside of yourself is the universe. There is no separation between you & universe, i.e., creation, as you are just as much creation as creation is creation.

So, by bringing yourself into mental harmony, with the identification of self as universe, *self as creation*, inseparable, inextricable, you must realize that since its genesis, the Universe has been in one eternal progression, as perfect as it ever was, yet perpetually and consistently evolving and progressing into the best as possible can be. The eternal evolution has been an everlasting state of perfecting, yet never "perfect[21]"

Do you understand the value & the weight of the consciousness, you have just been endowed with? The fact that although many claim perfection doesn't exist, and that God is only perfect, which may be true, you as a product, in creation of the creator of all creations are inseparable, and inextricably linked to the creation of the creator of all creations, as that is exactly what you are. So, this would only allow us to logically deduce that, since the universe has always been on a perpetual, eternal, progressive realization of itself, for itself, eternally perfecting never perfect, that is to say, always becoming as best as possible to be, you must begin to enter into the consciousness that you must begin & always stay on the path of eternal progression.

Eternal progression is the inherent nature and will of the entire universe. If you cease to progress at any moment, you will be swallowed by the whole & as long as you forever strive towards a higher degree of self, that is to say, to always be in the

---

[21] Here we are insinuating perfect as meant to convey "complete"

process of perfecting, which is to be in the state of perfection, which is to always remain perfect, you may aspire and achieve greater things than you have hitherto attained.

Furthermore, it is the lazy, the uninspiring, and those who choose to suffer by their own selective will & volition, who, due to their choice, render themselves impotent in this beautiful existence that we inhabit & never attain the establishment of *perfection as a way of being*.

What you must always & forever recognize is that, you must never mentally-subscribe to the notion that you are never (or are not) perfect. In fact, you were born perfect, you were born a self-perfecting mechanism. The more you come into harmony with the actuality of your fuller possibilities and thus free yourself from subjective realities of the limited conceptions of personal potential, the more you will begin to reap the greatest benefits and rewards of your own life. So, it may be true that nothing in itself is actually perfect *per se*. But everything is perfect at the same time. You now ask *"How can that be?"*.

Here is the last & final missing puzzle piece for the understanding of yourself as perfecting & perfect. Now that you know that you are a creation of the creator of all creations, which irrefutably means that you are just as much the universe, in fact, *that you are the universe*, never have not been the universe and that the universe is a creation which is eternally progressing under constant change, adapting new characteristics as well as traits, hitherto unseen, moving towards what can best possible be. What you must recognize by these facts & what most fail to grasp due to semantic dissonance, which is created by the many interpretations of the word "perfect", is that many simply imply & grant perfect a synonymous edge with the idea of "finality". For our purposes, *perfection ≠ finality*.

One may say that the creator of all creations themselves, that which precedes creation, is whole, complete & perfect in themselves. But as a *creation* of the creator of all creations in the existing in & as the existing eternal progression that is the universe, one must recognize that you are perfecting never perfect, continual never final. In this context, when we say "never perfect", what is meant is that, we are implying that there is no finality, no *punto finale*. So, you must recognize that you are perfecting, ever becoming and eternally progressing to your "best as possible can be *self*".

Still never will there be a moment of finality and definitive perfection, as the universe in & of itself is known to be 13.8 billion years old & still ever becoming. It is amazing that some of our minds can even comprehend this number. But it is incomprehensible beyond belief how insignificant that number (13,800,000,000) truly is. Yet to the human mind, which only lives on a scale of perhaps upwards the potential for 70-100 years, for most minds, 13,800,000,000 as a time figure seems like a number beyond digestion.

You are continually perfecting never finally perfect because the universe has always been perfecting & never has been perfect[22]. So as an inseparable, inextricable, actual piece of the universe, you must recognize that you have no point of finality in your eternal progression. Once you free yourself from the concept of a final point of progression, you've realized the process of being & becoming is all which you need to attach yourself to. Never final results or a final product. You are never a final product, you are always perfecting, never perfect.

We must transcend the crass usage of the word "perfect". You must now recognize that this approach in itself, which is the coming into harmony with a perpetual state of becoming "what is best possible to be" is what it truly means to be or become perfect.

The creator of all creations has created all things perfect by default. The only thing which is imperfect (or not perfect), that is to say, the only thing which is born defective is the individual human mind. There is no flaw or error in the universe, only in individual human thinking. Wisdom & Right Will are the only answers to life, for they allow us to allow ourselves, or if absent, disallow ourselves the best possible potential. God has willed that you have a will and you have willed everything into your life up until this point. Now that you have been availed & opportuned these instructions you may come into conscious harmony with these facts, if you so choose/will. If you are wise,

---

[22] Here perfect meaning it has never has reached a point of finality

you will consistently allow the unfolding process of self-perfecting evolution, being & becoming, to become autonomic.

You as a now consciously self-perfecting being will always be aspiring to your eternal progressive state, never ending or seeking a point of finality. This is because the universe in itself is in constant, unceasing, continuous motion & vibration. So, if you are to cease, you are to begin to die and to perish. This is cosmic law. It can & will only be by your comprehension and mental integration (assimilation) of these here facts which will then allow you to best live a life, mentally enlightened, so that you may most auspiciously progress on planes of existence.

Never allow anyone to allow you to believe that nothing is perfect. These individuals are the low aspiring souls, those with low aims and low performative desires. Those who do not want to assume the full responsibility that will grant them the fullest bounty of their lives. Those who will perish by their own choice & live less than what the universe has truly intended for them.

You are perfect, you were born perfect, you have never not been perfect & in fact, you were created a self-perfecting creation, just as the universe was created as a self-perfecting creation. The one thing about all things in existence is simply that you as a human, have free will, free choice of conduct. All other things in existence, must simply self-perfect & can never deviate from the law of their best being.

You possess the power to choose to abdicate or fully initiate this responsibility of best becoming. So now you have been endowed with the attitude of perfection on a cosmic scale, recognize that it is now your eternal responsibility and duty to have all the desired required characteristics as well as traits that your soul demands of you, which is truly the higher calling or the divine insistence.

From here onward, make yourself as great as possible can be and realize that, this standard will always forever be changing. The more you become, the more you can become, the more you will be, the more you can become, *ad infinitum*. The linguistic defects of English as a language will confuse many, but you are perfecting never perfect, meaning that there is no finality to your perfection. Yet remember that you always were & actually always have been perfect since your inception, from before you were born in flesh & were borne in the mind of the creator of all creations. Recognize & realize that you were designed as a self-perfecting being. Use your consciousness and your will to further refine yourself into that which your soul insists that you become and live to possess all the desirable characteristics traits, as best as possible to be. Live in the state of perpetual eternally progression.

Your world is Yours, always has been, never will not be.
- Eiman "Sheikh" Makki Completed October, 2023

BEFORE YOU WERE A MAN/WOMAN

∞

BEFORE YOU WERE
BLACK/WHITE/BROWN/OR ANY RACE
BEFORE YOU WERE HUMAN
BEFORE YOU WERE AS YOU ARE

∞

YOU WERE BORNE IN THAT
WHICH WAS BEFORE ALL THINGS EXISTED

∞

YOU ARE THAT WHICH
HAS EXISTED SINCE ALL THINGS EXISTED

∞

WHICH STILL IS & WHICH WILL BE AFTER ALL
THINGS EXIST AS THEY ARE

∞

YOU WERE IN "IT"
YOU CAME FROM "IT"
"YOU ARE "IT"

∞

YOU ARE THAT WHICH HAD A NAME
BEFORE THE FORMATION OF THE HUMAN BRAIN

∞

THE TRUTH HAS ALWAYS
BEEN HIDDDEN IN PLAIN SIGHT
The Cause Remains Laden Within The Effect

الله اكبر و إنَّا لله وَإنَّ إِلَيْهِ رَاجِعُونَ

Made in the USA
Middletown, DE
08 August 2025